This book is dedicated to the teachers of Rock Hill Schools.
Thank you for allowing me to serve you.

What Are Fungi?

Front and back cover photo by Carlton B. Griffith of Carlton Griffith Photography, Belton, SC.

All prose and poetry Copyright © 2015 by Kimberly Griffith Massey
(formerly Kimberly Griffith Anderson)

This book was first printed with funding from The Rock Hill Schools' Foundation, February 2016.

ISBN-13: 978-1519562562

Pronunciation from http://dictionary.reference.com/browse/fungi Accessed September 17, 2015

Artwork and layout by Kimberly Griffith Massey.

All rights reserved. No part of this publication may be reproduced, stored in a retrieval system, or transmitted in any form or by any means—electronic, mechanical, photocopying, recording, or otherwise—without the prior written permission from the author. The blackline master page at the end is an exception and is labeled as such. The blackline master is not to be shared or used by persons or entities who did not purchase an original copy of this book.

PHOTO CREDITS:
Cover photo by Carton B. Griffith of Carlton Griffith Photography – Belton, SC
Pages: ii, 1, 5, 8, 21, 26: by Carton B. Griffith of Carlton Griffith Photography – Belton, SC
Pages: 6, 19: by Deitrick Q. Sims – Charleston, SC
Pages: 2, 3, 4, 7, 9, 10, 11, 13, 14, 15, 17, 18, 20, 22, 25,
       28: top left, bottom left, bottom right by Kimberly Griffith Massey – Rock Hill, SC
Page: 28: top center, top right, bottom center by Ben E. Griffith – Sumter, SC
Page: 31: Author photo by: Nicholas F. Anderson – Rock Hill, SC
Page: 31: Photographer photo by Carton B. Griffith of Carlton Griffith Photography – Belton, SC

Contents

Introduction to Fungi . . . 1 - 6

Mushrooms . . . 7 - 15

Importance of Fungi . . . 16 - 19

Mold . . . 20 – 21

Yeast . . . 22 – 23

Athlete's Foot . . . 24 – 25

Summary . . . 26 – 27

More Pictures . . . 28

Glossary . . . 29

Reproducible Page

Pronunciation:
Fuhn-jahy
OR
Fuhng-gahy

Fungi are a unique type of living thing.

Teacher, what are fungi?
I've heard they're all around me.
Are they a type of plant?
Are they are a type of ant?
Are they green?
Are they mean?

No they're not plants,
And they certainly aren't ants!
But you might find them in the grass,
Or on other things you might pass.
Mushrooms are a common type of fungi.
If you find one in the grass, leave it be!
Some fungi are poisonous,
Others are helpful to us.
Fungi are neither plants nor animals, you see,
These living things have their own special category.

Okay, so mushrooms are fungi!
Is that all there is to know?

Well, yeasts are fungi, and so are molds.
With all the species, there are a million times five.
And like plants and animals, they too are alive!

Fungi are not plants.

Fungi do not have roots, stems, leaves or flowers. Fungi do not make their own food from sunlight. Some fungi, however, do grow up from the soil.

Fungi are not animals.

Fungi do not have arms, legs or wings. They cannot move on their own. As you will learn, however, some fungi have gills!

Keep reading to find out what fungi are and why they are important.

Have you ever seen a mushroom? A mushroom is a type of fungus.

The word **fungus** is used when there is only one. The word **fungi** is used when there are many. The study of fungi is called **mycology**.

Mushrooms often grow after it rains, and they grow up fast!

The mushroom is only a small part of the larger fungus that is growing underground. When a mushroom springs up, it means the fungus is ready to reproduce.

Mushrooms hold the fungal <u>spores</u>.

When the wind blows, the fungal spores are carried away from the <u>gills</u> of the mushroom to a new place to grow.

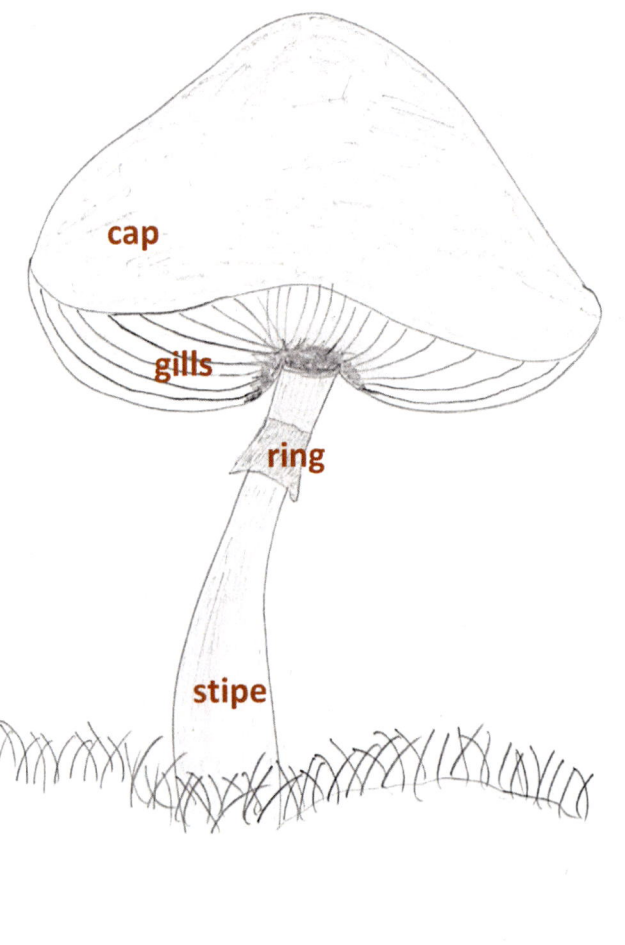

cap
gills
ring
stipe

Can you find the *gills* of this mushroom?

The spores of a fungus are like the seeds of a plant.

When the spores land, they will grow into a new fungus.

Mushrooms bought in the grocery store can be a tasty addition to many foods.

Never eat mushrooms found growing outside. They may be poisonous.

Can you count the mushrooms on this salad?

Can you find the mushrooms on this pizza?

For many years, fungi were thought to be plants.

Fungi, such as mushrooms, can often be found growing from the soil along with plants.

Plants are green because they make their own food.

Fungi do not make their own food and they are not green. Fungi are often considered to be <u>decomposers</u> because they break down living things after they die.

The job of a fungus is very important. Without fungi our world would be filled with dead plants and animals. Fungi help dead things to <u>decay</u> or <u>rot</u>.

Look at the fungi growing on these dead trees.

Another place you may find fungi is in your home!

Have you ever seen
a fuzzy slice of bread?
This type of fungus is called
<u>mold</u>.

Mold grows on food and many other things after they become old.

Amazingly, a fungus is used to make bread, too! The type of fungus used to make bread is called <u>yeast</u>!

A type of fungus called yeast is used to make bread rise.

When bread is made, flour, sugar, water and yeast are mixed.

The yeasts eat the sugar which makes the bread rise.

Baking the bread kills the yeast and leaves delicious bread!

One more place fungi can be found is on the human body!

Yes, fungi can grow on human skin.

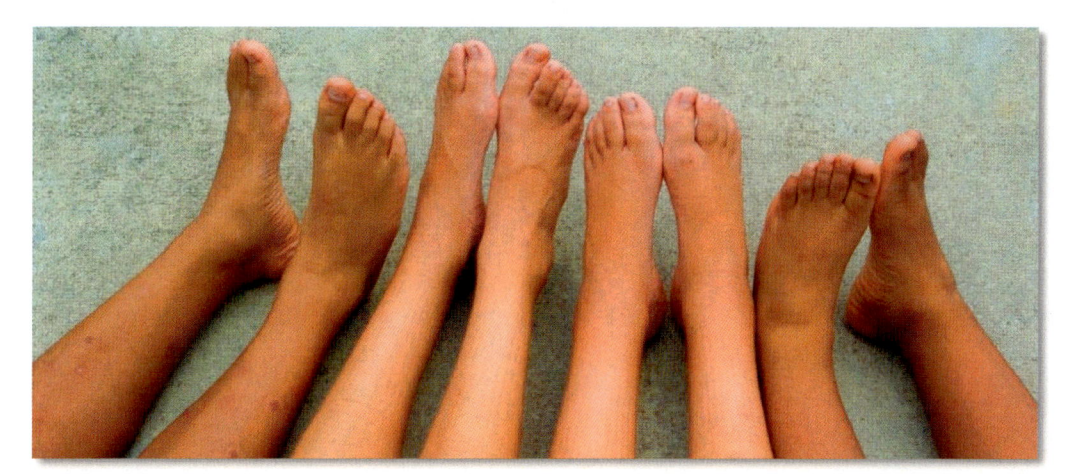

An example is when people have itchy feet due to the fungus that causes Athlete's Foot.

Keeping your feet, shoes, and socks both clean and dry can prevent Athlete's Foot.

You have learned about a unique type of living thing called fungi.

Summary:
- Fungi are not green and they do not make their own food.
- <u>Mushrooms</u>, <u>molds</u>, and <u>yeasts</u> are all different types of fungi.
- Many fungi are <u>decomposers</u>. They help dead things <u>decay</u>.
- Without fungi, there would be dead plants and animals everywhere.
- Without fungi, we would not have delicious bread.

Look for these fungi on your next nature walk.

Lichen is growing on this tree. Lichen is an organism made of algae and fungi.

28

Glossary

Athlete's Foot: a medical condition that is caused by a fungus that grows on the feet

Fungi: a group of living things that commonly help dead things to rot. Fungi were once thought to be plants, but are now known as a separate category of living things.

Fungus: a single organism in the fungi category

Decay: to rot or break down after death (Food will rot when it is old.)

Decomposers: living things that help dead things to rot

Gills: the area beneath the cap of a mushroom where fungal spores are held until they are released

Mold: a type of fungus that commonly grows on old, rotting food

Mushrooms: the part of a fungus growing in soil that appears when the fungus is ready to reproduce

Mycology: the scientific study of fungi

Rot: when living things (or foods) die and break down

Spores: the reproductive structures of a fungus (like the seeds of a plant)

Yeast: a type of fungus that can be used to make bread

What Are Fungi?

1. Fungi are not _____ or _____.

2. Three types of fungi are _____, _____, and _____.

3. Fungi are important because they are _____.

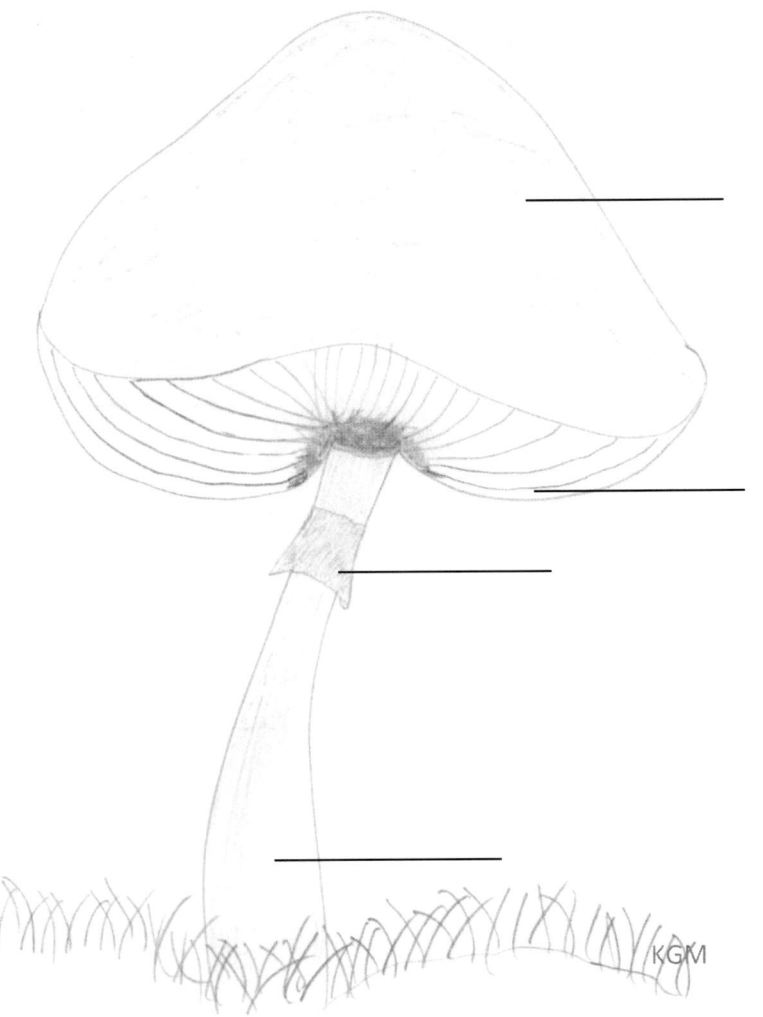

Color the mushroom, then label its parts.

1. plants, animals 2. mushrooms, molds, yeasts 3. decomposers See p. 10

This page is reproducible.

Stay Connected:

Email the author at goodgirlnovel@yahoo.com

   Books by Kimberly Griffith Anderson

   @kgandersonbooks

   @kgandersonbooks

Use **#fungifun** to show your latest and greatest fungi pictures!

Other books by Kimberly Griffith Anderson are:
- Good Girl
- Single Dad 19
- But I Love My Husband
- But We're Not Married

All are available at www.amazon.com

To see more work by Carlton Griffith Photography visit:
www.carltongriffith.com

The Author

Kimberly Griffith Anderson is a Science educator from South Carolina who is educated in the Biological Sciences. She enjoys teaching students of all ages about various aspects of nature. This is her first children's book, but she has previously published four novels. She is a proud wife and mother whose primary pastime is writing.

The Photographer

Carlton B. Griffith is a portrait and fine art photographer based in beautiful Anderson County, South Carolina. He loves capturing and sharing the beauty of God's creations, especially human faces and the many wonders of nature. He is also a husband and father, an elementary school tech guy, and a church sound guy.